Before Stuxnet

ROMAN POROSHYN

Copyright © 2017 Roman Poroshyn

All rights reserved.

ISBN: 1979882150
ISBN-13: 978-1979882156

This book may not be reproduced, transmitted, or stored in whole or in part by any means, including graphic, electronic, or mechanical without the express written consent of the publisher except in the case of brief quotations embodied in critical articles and reviews.

CONTENTS

The Back Door to the Nuclear Powers Club 1

A. Q. Khan and the North Korean
Nuclear Bomb 11

Gernot Zippe – the Creator
of Uranium Enrichment Centrifuges 21

Flashback: German Researchers
in Soviet Captivity 33

Sabotage –the Weapon of Choice 41

Glossary 49

Selected Bibliography 61

CHAPTER 1

The Back Door to the Nuclear Powers Club

Before 2010, the idea of creating a computer virus capable of attacking uranium enrichment facilities would have sounded crazy to the majority of people. Those attacks would have been impossible to implement because uranium enrichment is one of the most crucial and heavy guarded industries for every nation-state. It has been unattainable for any computer virus to penetrate the various levels of defenses including firewalls, anti-virus software, and an air-gap, which effectively cuts the facility off from the Internet.

After the discovery of Stuxnet, we could not underestimate the power of malware. Now we know for sure that It can find its way through an air-gap and protective measures. It is possible. Even worse, it was implemented. The computer worm Stuxnet was able to reach its target and attack Iranian uranium enrichment centrifuges five times during the period from June 2009 to May 2010.

Years later, the amount of preparation and groundwork involved in the preparation of Stuxnet still captures one's imagination. It was much more involved than just sophisticated programming code. It required detailed knowledge about processes and equipment used, uranium enrichment centrifuges, in particular, a configuration of cascades, etc. It means some solid intelligence gathering. Of course, for this reason, the malware Duqu has been created. Still having Duqu snooping inside computer networks would not be enough. Having an actual physical sample of a centrifuge would be helpful too.

At one point in time all the factors needed for the creation of Stuxnet miraculously came together. It all began with two isolated events, which happened in different geographical places and years apart from each other. The first event took place in India near Pakistan's eastern border in 1974. The second event happened in the coastal city of Taranto, Italy, in 2003. Despite the fact that both events took place outside Iran, they combined to create a background for everything that later would become known as the Stuxnet computer worm and its cyber-attacks against the Iranian uranium enrichment program. For this reason alone, those events and their aftermath deserve a closer look.

On May 18, 1974, India conducted a successful test of a nuclear device. It came as an unexpected and bitter surprise for India's neighbor, Pakistan. Only three years had passed since the end of the India-Pakistan Winter war, which ended with Pakistan receiving a humiliating defeat of its army and a loss of a territory (East Pakistan became a new independent state - Bangladesh). Now Indian nuclear test announced to Pakistan that its rival was well on its way to possess its very own nuclear weapon. The underground test took place at the Pokhran Test Range located in the Thar Desert (a.k.a. the Great Indian Desert) near the border with Pakistan. This test, which became known under the codename "Smiling Buddha", announced to the world that India had become the newest member of the Nuclear Powers club. At the same time, the echo of this test led to some unintended and unexpected consequences. One of them was widespread nuclear proliferation around the world, but India is not the one the one to blame.

Pakistan was not a novice in seeking ways to build an atomic bomb. It was working on its own atomic bomb project since 1972, but no sufficient progress toward having a bomb was accomplished. There were two parallel projects aimed at the creation of a nuclear bomb. One project was based on using plutonium, and the other was focused on utilizing uranium for a bomb. Things changed drastically after India's nuclear test in 1974. At that time, the Prime Minister of Pakistan was Zulficar Ali Bhutto. Back in 1965, he made the following statement, "If India builds the bomb, we will eat grass or leaves, even go hungry, but we will get one of our own. We have no other choice." Now, almost ten years later, it was the time for Bhutto to bring his old promise to life. He could not fulfill the promise on his own, so he needed help.

The possibility of help had materialized on September 17, 1974, when Dr. Abdul Qadeer Khan, who worked for Urenco Group on its weapon-grade centrifuge production facility in the Netherlands, wrote a letter to Prime Minister Bhutto. In his letter, Dr. Khan offered his help in establishing the production of enriched uranium in Pakistan. Soon he returned to his home country loaded with stolen blueprints and other technical documentation. After four years of trials and errors, the "made in Pakistan" centrifuges produced their very first enriched uranium.

      Dr. Khan quickly advanced from being in charge strictly for the uranium enrichment to a position, which allowed him to exercise unlimited control over all aspects of Pakistan's nuclear program, including a creation of a nuclear weapon. As his influence grew, Abdul Qadeer Khan became an indispensable and uncontrollable. Known as "the father of Pakistan's bomb", at one point, he made a decision to become "godfather" of other nations' bombs. Dr. Khan was ready to sell the nuclear technology to others. He had everything for that – proven success record, technical documentation, and a vast network of suppliers. Now he needed to find who would be willing to enter the Nuclear Powers Club through the back door. For them, Dr. Khan was willing to sell the keys to that door.

      Not many details are known about how the first uranium enrichment centrifuges, decommissioned in Pakistan, ended up in Iran,

Pakistan's western neighbor, but this is how the Iranian uranium enrichment program received its jump-start. What is known is that the delivery of centrifuges together with related technical documentation was the direct result of actions by the international nuclear proliferation network commonly known as the "Khan's network". The public exposure of the Khan's network came in December 2003, after Libya gave up its nuclear weapons-related supplies. Among other materials sold and delivered to Libya by the Khan's network were uranium enrichment centrifuges and blueprints of a nuclear weapon designed by China.

    The Central Intelligence Agency (CIA) knew about the existence of this nuclear proliferation network for decades. Dr. Khan was on the U.S. intelligence watch list since the time he worked in the Netherlands. In 1979, the CIA analysts concluded that Pakistan had accumulated almost everything needed to build its own uranium enrichment plant. The CIA attempted to block some exports of equipment from Europe to Pakistan, but it was too little too late. The pace of the uranium enrichment in Pakistan was picking up, and the CIA began to look at other options. Eventually, it succeeded in recruiting some Swiss-nationals, who were prominent members of the Khan's network, to be its paid informants. One of them even became involved in CIA's efforts to sabotage the uranium enrichment equipment produced by Khan's network.

After years and years of watching and observing activities of the network, the CIA was ready to implement its first active action against the network in 2003. The plan was to locate and intercept a shipment of uranium enrichment centrifuges and use it as proof of nuclear proliferation. The opportunity presented itself when the CIA received a tip from its informant that a load of centrifuges would be delivered from Malaysia to Dubai and then to Libya.

The German-owned ship BBC China was on its way from Dubai to Libya when it suddenly changed its course after leaving the Suez Canal.

The ship's captain received the order from his boss in Hamburg, Germany, to go to the Italian port, Taranto. There was no explanation offered to the captain. He just followed the order. When, on October 4, 2003, BBC China entered the Italian port, the ship was lead away from the commercial part of the port to one of the docks under control of the Italian Navy. A group of uniformed Italian and US naval officers accompanied by several civilians was waiting for the ship. Even though they knew in advance the numbers of five shipping crates of particular interest, it took them several hours to find crates on the cargo ship and unload. After that delay, which nobody clarified, the captain returned the ship to its original course toward its final destination – Libya. Inside of those five crates, the agents of the CIA found what they expected to find – parts of uranium enrichment centrifuges. The good news was immediately reported to the CIA headquarters in Langley, Virginia.

Made in Malaysia at the Khan's network

machine shop, those centrifuges were a small part of the complete uranium enrichment facility supplied to Libya. Never delivered centrifuges served as a proof of Libya's involvement in nuclear proliferation. Under the diplomatic pressure, Colonel Gaddafi, who was the leader of the country at the time, decided to give up all equipment and technical documentation received from Khan in order to avoid new economic sanctions against Libya.

The long journey of the uranium enrichment centrifuges led them from Malaysia to Libya to the USA to Israel, where they were utilized to test the infamous computer worm Stuxnet. It is a well-known fact that Stuxnet was used to attack Iran. There have been some speculations that a similar cyber-attack was launched against the North Korean nuclear program because the same uranium enrichment centrifuges were used at the Iranian nuclear program.

CHAPTER 2
A. Q. Khan and the North Korean Nuclear Bomb

Even before becoming highly praised in his own country as "the father of Pakistan's atomic bomb", Dr. Abdul Qadeer Khan was a well-known figure in Western intelligence circles. After his sudden return to Pakistan from Europe with drawings and specifications for uranium enrichment centrifuges, his activities were closely watched. It turned out to be a very long watch. A. Q. Khan had been observed touring Europe purchasing dual-use equipment. The sellers of that equipment knew well enough that it could be used for uranium enrichment, but they sold it anyway since there was no ban at that time on selling that equipment to Pakistan.

All European organizations in charge of nuclear non-proliferation kept watching. From time to time sales of certain types of equipment were blocked. In those situations, Khan would switch to buying components and spare parts to assemble the prohibited equipment directly in Pakistan. Back then, the prevailing opinion was that the third world countries would be unable to perfect a complex and technologically advanced process as uranium enrichment. A. Q. Khan did everything in his power to prove that opinion wrong, and he succeeded.

After establishing uranium enrichment technology in Pakistan, Dr. Abdul Qadeer Khan kept buying equipment and spare parts in quantities much higher than Pakistan's own internal needs. The scale of those activities puzzled Western intelligence analytics. Again, nobody wanted to believe that some nuclear proliferation activities were taking place. At that time, the United States was involved in the covert war with the Soviet Union in Afghanistan. The support of Pakistan was vital for the US during ten years of this war. The US presidents were closing their eyes on Pakistan's progress in the nuclear weapons program. They kept certifying before the Congress that Pakistan did not have an atomic bomb despite the fact that reports by the CIA were stating the exact opposite.

With the help of A. Q. Khan's network, the Pakistani version of Zippe-type uranium enrichment centrifuges started to spread around the world. Those movements were highly classified and vigorously protected. For this reason, CIA and British Secret Intelligence Service (MI6) tried harder than ever to penetrate Khan's network. As always, when it comes to cloak-and-dagger operations, there were plenty of non-confirmed stories and conspiracy theories. According to one story, in June 1998, the samples of uranium centrifuges P-1 and P-2 were secretly delivered to North Korea by air under the official cover of returning home the body of a Korean diplomat's wife, who tragically died in Pakistan's capital Islamabad.

The name of the woman was Kim Sa Nae. On June 7, 1998, she was shot dead in an affluent neighborhood of Islamabad. The rumors were that it happened in A. Q. Khan's guesthouse, where an official delegation from North Korea was staying. The official investigation quickly ruled out that it was an accidental shooting. A neighbors' cook mishandled a shotgun borrowed from a security guard, which led to a fatal shot being fired.

That was the official version of events, which could not explain why neighbors heard more than one gunshot fired and the fact that Kim was shot at point-blank range. There was no autopsy performed and no case was registered by the police. Afterwards, the local newspaper published a four-line story about the mysterious murder of a North Korean diplomat. This story did not make headlines around the world because the world was busy watching another story unfolding – the nuclear arms race between India and Pakistan.

Historically the timing of nuclear tests is determined more by political interests rather than scientific needs, and the nuclear tests show-off between two rival countries, India and Pakistan, was not an exception. Twenty four years passed since India performed its first nuclear test until India's leadership decided that a nuclear test would be an appropriate response on Pakistan's test of a Ghauri ballistic missile. One month later after the missile's test, on May 11, 1998, India exploded three nuclear devices at the Pokhran underground testing site near the Pakistan border. Two more explosions followed two days later, on May 13, 1998. Nuclear tests cannot be prepared in a hurry. It takes months and months of hard work. Still, Pakistan was fast to respond with series of nuclear tests of its own. On May 28, 1998, Pakistan detonated five nuclear devices at Ras Koh Hills in the Chagai District of Balochistan. One more test was performed on May 30, 1998.

The Observation Post for Pakistan's test was established at about six miles distance from the mountain, stuffed with nuclear devices. Twenty people, including government and military officials, scientists, and special guests watched how at first smoke and dust came from the five sealed entrances to tunnels hiding nuclear explosives. Then the mountain itself shook and changed color before it disappeared from sight completely covered by a cloud of dust. Among the guest- observers were some North Koreans, possibly Kim Sa Nae, who would die ten days later, and her husband, Kang Thae Yun. The main reason for them being there was that they had pretended to be a low-profile North Korean diplomat's family. Kim Sa Nae appeared for outsiders as a housewife well-known among foreign diplomats for her cooking skills, especially for cold Korean buckwheat noodles. She turned out to be a North Korean nuclear scientist. Her husband, Kang Thae Yun was officially an economic counselor in the embassy, no more than an average paper-pusher. In reality, he was a nation-scale arms dealer, representing interests of North Korean political and military elite.

The existence of a mutual interest between North Korea and Pakistan was obvious. A nuclear bomb alone is almost useless without means to deliver it. Pakistan had a bomb and North Korea had a ballistic missile. Both countries were cash-strapped, so the barter deal was the ideal solution. The deal was arranged from Pakistan's side by A. Q. Khan, and Kang Thae Yun arrived in Pakistan as the representative of the North Korean side. As for Kim Sa Nae, she was in Pakistan as a part of twenty people team of scientists, who came to learn more about the enrichment of uranium by using centrifuge technology.

Some journalists, including Dexter Filkins (then from The Los Angeles Times) and Simon Henderson (from Foreign Policy magazine), shared the opinion that Kim Sa Nae, the wife of the North Korean diplomat, came under the suspicion of supplying Western intelligence agencies with information regarding uranium enrichment cooperation between her country and Pakistan. After that, even a diplomatic immunity could not help her to stay alive. The majority of observers seem to agree with that despite some differences into accounts describing the return of Kim's remains home.

According to Gordon Corera, the Security Correspondent for BBC, in order to bring the body of the diplomat's wife back home, the Pakistan government arranged for a Pakistan Air Force Boeing 707. At the last moment before the departure, Dr. Abdul Qadeer Khan himself showed up. He brought on board five large heavy wooden crates. Apparently inside the crates were uranium enrichment centrifuges P-1 and P-2 alongside with technical documentation, and possibly even a container with uranium hexafluoride for testing centrifuges. Nobody even tried to check Khan's luggage, because of A.Q. Khan's special status in Pakistan. He enjoyed complete freedom of crossing the borders. His organization had a virtually unlimited budget, and he reported his actions only to the president of the country.

At the same time, as Western journalists seem to be in agreement about the content of Khan's belongings, they disagree about the plane. Paul Watson and Mubashir Zaidi, in their article in The Los Angeles Times, stated that the airplane was not a Boeing 707, but actually a military cargo transport C-130. Also, not everybody agrees with the statement that Kim Sa Nae herself was on board of the plane heading toward North Korea.

Korea Web Weekly, for example, has been questioning even the sole fact of Kim Sa Nae's very existence. Its doubts were based on facts that there was no autopsy performed and no case opened by local police. Korea Web Weekly came up with its own version of events. No dead body meant no murder took place, and the whole story was a cover-up to distract CIA's attention from the withdrawal of twenty North Korean's nuclear scientists from Pakistan on the Air Koryo (the North Korean airline) plane loaded with nuclear test equipment and data obtained during the second nuclear test, which was in reality the detonation of the North Korean made nuclear device.

    At the end, it does not matter, which conspiracy theory sounds more convincing. The reality is that the Pakistan's uranium enrichment centrifuges made their way to North Korea and contributed to the creation of another nuclear weapon state because all the opportunities to prevent that from happening had been neglected at first and eventually missed.

CHAPTER 3

Gernot Zippe – the Creator of Uranium Enrichment Centrifuges

After the Khan's nuclear proliferation network had been exposed, the world's attention turned to the creator of now infamous centrifuges – Gernot Zippe. He became somewhat of a celebrity. Previously known only to a narrow circle of uranium enrichment experts, now Zippe was giving interviews to journalists from popular newspapers and magazines. He could be heard on the radio too. The new life in the spotlight was supposed to be a pleasant experience, but not for Gernot Zippe.

Every time he made a public appearance, Zippe needed to explain himself, and even protect himself against allegations. He was accused of several things. First, of helping Soviets to build an atomic bomb. Second, in simplifying uranium enrichment technology. In the eyes of the general public, Zippe made the step from the "power plant ready" low-enriched uranium to "weapon-grade" highly enriched uranium just a question of time (by adding additional cascades of centrifuges and spinning them for longer times the same uranium enrichment centrifuges would be able to produce a "bomb-ready" final product).

In 2004, BBC named its radio interview with Gernot Zippe "The Zippe Type: Poor Man's Bomb". The same year, The New York Times published the article about him under the title "Slender and Elegant, It Fuels the Bomb". Apparently, some kind of repentance was expected from Dr. Zippe, but he did not budge. His most often repeated quote had been the following, "With a kitchen knife you can peel a potato or kill your neighbor. It's up to governments to use the centrifuge for the benefit of mankind." William J. Broad from The New York Times commented on that, "If a chief inventor [Gernot Zippe] has any regrets, he keeps them private." Apparently, there were no regrets at all. In his less known statement, Zippe said, "Gas centrifuges... made an important contribution to the fact of Cold War between East and West came to an end without a nuclear holocaust".

The role of uranium enrichment centrifuges in the peaceful resolution of the Cold War, as Gernot Zippe described it, could be arguable. Still, the fact remains that Zippe-type centrifuges have opened the door to the Nuclear Powers Club to many nation-states. For this reason alone, gas centrifuges deserve a closer look.

In its natural form, uranium ore is a mix of three isotopes of Uranium. One of them is the highly desired Uranium-235, which makes less than 1% of the total mix. Because of Uranium-

235's unique ability to support a nuclear chain reaction, it is suitable for use as fuel for nuclear power plants or as the main component of a nuclear warhead. Another, the most common isotope is Uranium- 238. It represents more than 99% of uranium ore. When it comes to the production of nuclear fuel, Uranium-238 is useless, because it cannot support a nuclear chain reaction. The other isotope is Uranium-234 (about 0.005%). This last isotope is nothing more than a decay product of Uranium-238. Since chemically all three isotopes are identical (they are essentially the same chemical element), it is impossible to separate them by crystallization, distillation, evaporation or any other technique commonly used for separation of different chemical elements.

The main difference in physical properties between the isotopes of Uranium is their weight. Uranium-235 is slightly lighter than Uranium-238, but how much lighter? The numbers in the names of isotopes describe their atomic weight (or scientifically speaking - the atomic mass).

Scientists have agreed to calculate the single unit of atomic weight by dividing the atomic weight of a single atom of Carbon by twelve. Since the result of this purely theoretical mathematical operation is virtually impossible to measure in ounces or even grams, there is no surprise that atomic weight has no dimension assigned to it. The non-dimensional weight difference between Uranium-235 and Uranium-238 is less than the weight of a single atom of Carbon-12. Scientists have been able to utilize this tiny difference in weight in order to separate Uranium-235 from Uranium-238.

The two commonly used techniques for separation of Uranium isotopes are gaseous diffusion and centrifugation. Both techniques use uranium hexafluoride as raw material. In the gaseous diffusion process, uranium hexafluoride is forced through a series of porous membranes. Under pressure, the light-weighted Uranium-235 moves through membranes faster than Uranium-238. After passing through several hundred membranes, the concentration of Uranium-235 is high enough to be used for fabrication of nuclear fuel.

Centrifugation has proven to be more energy-efficient. It uses significantly less electricity (50 kWh compared to 2,500 kWh per Separative Work Unit – unit of measure for enrichment process). In addition, centrifugation is proven to be a more discrete process (from outside a building that houses uranium enrichment centrifuges would look like an ordinary warehouse or machine shop) compared with a much larger and easily distinguishable footprint of a gaseous diffusion plant.

By definition, a gas centrifuge works with uranium in its gaseous form – uranium hexafluoride. It takes time and many centrifuges, connected with each other in cascades, to enrich uranium. By itself, the concept of uranium enrichment by using centrifuges looks seemingly simple. Its implementation, as the history of uranium enrichment centrifuge demonstrates, has been a long and difficult process with some unexpected twists.

The idea of using centripetal force for separation of isotopes with different weights, the process is known as centrifugation, was proposed back in 1919. Since it is easier to have things said than done, it took another fifteen years of failed attempts to prove that this idea would work. In 1934, Jesse Beams (1898-1977), a professor of physics at the University of Virginia, successfully separated isotopes of chlorine using a centrifuge. Beams came up with one of the key elements of the contemporary centrifuge design. He did that by placing the rotating part of the centrifuge (its rotor) inside a stationary cylindrical housing and made it spin in a vacuum. The other elements of the contemporary centrifuge design were developed in the 1950s in the Soviet Union. It was not an easy task to pull them out from behind the Iron Curtain, but eventually, it happened.

The centrifuge research in the Soviet Union was originally led by Fritz Lange (1899-1987), who escaped from Nazi's Germany in

1935. The first Soviet uranium enrichment centrifuge was designed by Lange in 1944. By today's standards, this Lange's centrifuge was slow (its rotation speed was only 18 meters (sixty feet) per second), short – only 600 mm (two feet) long, and with very thick walls. The testing of this centrifuge design confirmed the original prediction: the desired level of enrichment cannot be accomplished within a single centrifuge. Somehow, uranium should be transferred between centrifuges in order to make the uranium enrichment a multi-stage process. Every stage would gradually increase a concentration of the desired Uranium-235.

After the end of World War II, more German researchers and engineers, who were prisoners of war, were brought in to work on classified projects in the Soviet Union. One of them was the world-known physicist Max Steenbeck (1904-1981). He became the head of one of the several groups that worked independently on developing a uranium enrichment centrifuge. Steenbeck explained the behavior of isotopes in a centrifuge during the separation process. His theoretical findings were immediately put into practical use. It was determined that in order to make the enrichment process more efficient both the length of the rotor and its speed of rotation needed to be increased significantly. That is why the initial efforts of Steenbeck's group have been focused on much taller (10 feet long) and faster (almost 800 feet per second) centrifuge than one designed by Lange. It was a significant improvement compared to the earlier design. Still, Steenbeck's centrifuge had two major flaws. First, it required a precision balancing of each individual rotor, which made it difficult to manufacture in industrial scale. Second, there was no solution yet for transferring gaseous uranium from one centrifuge to another.

The breakthrough came in 1952, when Evgeni Kamenev, from the group under the lead of Isaak Kikoin (1908-1984), shared with colleagues his concept of a short (about one foot long) centrifuge that would be able to achieve the speed of one thousand feet per second. Kamenev's design incorporated the original ideas of his group (easy to produce short unbendable rotor and ability to transfer gas between centrifuges) with ideas of Steenbeck's engineers (needlepoint bottom bearing and magnetic bearing on the top). Also, Kamenev reinvented Jesse Beams' idea of spinning the rotor in a vacuum in order to reduce friction and achieve higher speed. The prototype of Kamenev's centrifuge with one and a half foot long rotor has been completed in December of 1953.

Earlier, in the summer of 1953, another talented engineer from Steenbeck's group and a prisoner of war, Gernot Zippe (1917-2008), built his own prototype with the even shorter rotor (only one foot long). From this point, the Soviets continued working on improvements to Kamenev's prototype for their uranium enrichment program, and captured Germans were removed from any classified research. After more than ten years of working in captivity, Zippe and other prisoners of war were freed.

Gernot Zippe returned to Vienna, Austria in 1956. After attending the International Symposium on Isotope Separation in Amsterdam in 1957, Gernot Zippe realized that the Soviet Union had been far ahead of the West in this particular area of research. Inspired by his advanced knowledge and experience, Zippe continued his work on enrichment centrifuges, first in Germany. Then in 1957, he crossed the ocean under fictitious documents, supplied by the CIA. In the USA, he ended up working with Jesse Beams at the University of Virginia, at the birthplace of the first enrichment centrifuge. Here, working under a contract with the Atomic Energy Commission, Zippe, by memory, recreated his prototype from 1953.

Despite the recruitment efforts by the CIA to convince Gernot Zippe to change his citizenship and stay in the USA, he returned to Europe, where, in 1960, together with his colleague from Soviet captivity - Rudolf Scheffel, he patented the centrifuge design. Nobody else from more than a hundred people, the majority of whom he knew personally, had been included as the authors of the invention. The patent had been granted, and almost immediately all research related to uranium enrichment centrifuges has been classified in Europe per request of the US government.

Since that time, this design, known as the Zippe-centrifuge, continues to be widely used around the world for uranium enrichment.

CHAPTER 4

Flashback: German Researchers in Soviet Captivity

As an aftermath of the World War II, Germany lost a whole generation of its physicists. Some were killed on battlefields, others ended up as prisoners of war, and the others were forced to immigrate to the USA or the Soviet Union, depending on in which zone of occupation they found themselves at the end of the war. Among other scientists, German physicists were treated by the victors as a highly valued war trophy, which should be put to use in their homeland.

One of the Soviet prisoners of war was Max Steenbeck. Because of his extensive background in physics research at Siemens, Soviet officials in charge of finding "trophy scientists" removed him from a prisoners' camp and appointed him as the head of a newly created research laboratory. As a sign of trust, Steenbeck was allowed to select his future subordinates from the list of German scientists held in Soviet prisoners' camps. One of Steenbeck's picks was Gernot Zippe, a former Luftwaffe flight instructor, gifted mechanical engineer, and future creator of the Zippe-type centrifuge.

Thanks to Max Steenbeck, after six months in a camp, Zippe found himself out of the camp and in the research institute "A", located near Sukhumi in Georgia, the Soviet Union. The letter "A" in the name of the research institute came from the last name of its German director Manfred von Ardenne. The official name of the institute "A" was "Enterprise P.O. Box

0908". During the times of the Soviet Union, it was a common practice to hide highly classified objects related to the national defense, such as factories, machine shops, and even research centers behind the nameless numbers of post office boxes.

Before World War II, there was the Sinop sanatorium. Now it became the territory of the institute "A". It was fenced and guarded by military personnel. German scientists and their families worked and lived there, behind the fence. At the same time, they were allowed to visit grocery stores and a restaurant in a nearby town, even go hiking. Whenever Germans stepped outside the fenced territory, they were escorted by an armed guard, dressed as a civilian.

An experimental centrifuge for uranium enrichment was located in the building "L". It was a two-story building with a long hallway that went from wall to wall parallel to the facade of the building. One of the rooms on the second floor had a hole through the floor to accommodate the whole height of the centrifuge. In this room, researchers were collecting enriched uranium hexafluoride into a removable glass ampoule. Doors to both rooms, which housed the centrifuge, were secured by checkpoints with armed guards. Guards were responsible for verification of ID with proper security clearance before letting anybody into the room. Despite the fact they knew everybody, who had access to those rooms, the guards continued checking IDs. Only five people were allowed to be there: Steenbeck, Zippe, and their three Soviet assistants.

The centrifuge was in operation around the clock, days and nights, stopping only for replacement of a failed rotor. Zippe personally did a rotor assembly, since a rotor was the most critical part of the whole experiment. Steenbeck and Zippe were continuously monitoring the results of uranium enrichment. Often Zippe would show up in the middle of the night to check the condition of the centrifuge and make sure his Soviet assistants were watching the process. The main objective was finding a way to stabilize a rotor, which could not survive the high speed of rotation for a prolonged period of time.

The accomplishment of this objective was a higher priority than the safe working conditions. After another rotor would tear itself apart, which was a common occurrence, often an evacuated glass ampoule would be shattered too, and uranium hexafluoride gas would fill the room. In this situation, researchers were instructed to turn on a fan and take a short break outside. The fan would force air from the room into the hallway. The security protocol required to keep all vapors inside the building. Despite the warnings of health risk from the researchers, a guard could not leave his post. He stood there the whole time inhaling poisonous fumes.

When the researchers came back, they would disassemble the stationary casing of the centrifuge and wash its parts in a metal tank filled with dichloroethane before reassembling them. Those parts had little balls of mercury from dampers and tiny particles of uranium on them. Researchers were in contact with mercury and uranium in addition to breathing toxic vapors of dichloroethane. There were no respirators or other personal protective equipment in use. Researchers received free milk, which was recognition of the presence of health hazard according to the Soviet occupational health standards of that time. Apparently, milk did not do much. Unidentified sicknesses with strange symptoms were common among the people involved in this particular research.

Zippe's life in the Soviet Union was surrounded by myths. The most widespread myth was that before Zippe's return to Austria, Soviets took away all his notes. In reality, there was nothing to take away. Zippe never had any work-related notes in his personal possession during that period of his life. He had his work journal to record the accumulated data. In the same journal, Zippe recorded his observations and new ideas to try. At the end of his workday, like every other employee at the institute "A", he would put his journal inside a binder and seal the binder with his personal seal. Then Zippe would lock his room (it would be a stretch of the imagination to call his workspace an office) and use his personal seal to secure the entrance into the room. After that, he would turn in his binder to an office of the First Department.

During the Soviet Union time, the First Department was in every organization. Its main responsibility was keeping secrets in secret. Once a week, people from the First Department would do a routine search. They would check every workstation looking for loose papers and notes. The mandatory rule of the security protocol was that everything must be written inside a work journal with numbered and threaded pages. Both ends of the thread were secured with the First Department's seal (so nobody could covertly take a page out or replace it). For this reason, there were no sketchy records in secret notepads, which could be taken away from Zippe.

Unfortunately, for the officers from the Soviet's First Department, Zippe kept everything in his head, and they knew that. They came up with a remedy called "quarantine". In order to give Zippe and others time to forget, all German researchers were removed from any classified work for more than two years before their departure, Gernot Zippe remembered how to build a centrifuge. He put his knowledge and expertise to use first in the USA, then in Western Europe.

CHAPTER 5

Sabotage –the Weapon of Choice

Sabotage had become the main weapon of choice for Western secret services in their secret war against Iranian nuclear program. Attempts to penetrate the supply chain of Khan's network and its customers with faulty parts and inferior materials was an invisible ongoing activity for years. Few of those attempts became known to the general public. Among them were the alteration of seven vacuum pumps build by the German company Pfeiffer Vacuum Technology and an unspecified quantity of Turkey-made uninterruptible power supplies (UPSes). The vacuum pumps were sold by the original manufacturer directly to the Los Alamos National Laboratory, where they were modified to make sure they would fail, and resold to buyers from Khan's network. As for the UPSes, they were modified during their journey from Turkey to Libya and Iran. More than likely, that happened in a remote warehouse in Dubai, where the CIA set up its field workshop.

Less known sabotage attempts included minor changes in drawings of centrifuges built by Khan's network in Malaysia. The changes varied from alterations of hole diameters to a replacement of required material from steel to aluminum. One of the changes was a removal of the requirement of cleaning parts of centrifuge rotors in an ultrasound bath before the final assembly. It has been proven that tiny dust particles, even drops of sweat from assembler's hands if left on the rotor, could be enough to throw it out of balance during the operation at high-speed. Obviously, all those sabotage attempts were thoughtfully designed to lead toward a premature failure of equipment involved in the uranium enrichment process. The main question was how successful those sabotage attempts actually were.

European experts from the Urenco (the company from which the original centrifuge design was stolen by A. Q. Khan), per the request of the International Atomic Energy Agency (IAEA), performed an evaluation of the samples of the sabotaged uranium enrichment centrifuges recovered from Libya. After testing them at their facility in the Netherlands, Urenco reported to the IAEA that the centrifuges were completely functional with possibly lesser operational life than expected. Apparently, there was a difference in opinions between European engineers from Urenco and "mad scientists" employed by the CIA.

Sabotage is usually justified as the best tool for making all efforts by a sabotaged party to accomplish the desired result fruitless. In reality, sabotage alone could only delay or slow progress, but never indefinitely prevent. Even more important is the sad fact that uncovered sabotage attempts have become a valuable learning experience for the sabotaged party. Iran has not been an exception. Iranian scientists and engineers, many of them Western-educated, proved to be quick learners. During the period of the work suspension at the Natanz uranium enrichment plant (end of 2003 – beginning of 2004), inspectors from IAEA confronted Iranian technicians doing some work on centrifuges. Iranians replied that they were doing repair work replacing failed O-rings. The made in Britain O-rings were, either intentionally or by mistake, manufactured from an inferior material that could not resist the amount of pressure building up during feeding and withdrawing of uranium hexafluoride. In this case, Iranian engineers successfully zero-downed to the reason for the failures and found the solution.

For some sabotage attempts against Iran, stakes were much higher, because they were focused on sabotaging the nuclear weapons program. A good example of such attempts has been Operation Merlin, which took place in 2000 and came back to the spotlight in 2015.

On May 11, 2015, Jeffrey Sterling was sentenced for releasing classified information to a reporter. Since his arrest in January 2011, Sterling was living under the threat of a long prison sentence, somewhere between nineteen and twenty-four years according to sentencing guidelines. Fortunately, for Sterling, the U.S. District Judge Leonie Brinkema has decided that "the guidelines are too high" and forty-two months punishment would be enough.

As a CIA officer of the Iran Task Force during the period from 1998 to May 2000, Sterling was a curator of a Russian immigrant with a background in nuclear science. Even today this immigrant is known only by his codename Merlin. Merlin's objective was to supply Iranians with drawings of Russian design of an important component of a nuclear weapon. This component, the TBA-480 Fire Set, is responsible for triggering an initial explosion necessary to start a nuclear chain reaction. The drawings were modified to assure that the component would fail.

This sensational story became known to James Risen, a reporter from The New York Times. At first, he wanted to do exactly what all reporters do: spread the news around. Surprisingly enough that was not an easy task to do even for The New York Times, which is well-known for whistle-blowing. Still, back in 2003, Condoleezza Rice, then-National Security Adviser, had a meeting in person with Jill Abramson, the Chief of the Washington Bureau of The New York Times, and James Risen, the reporter. Rice successfully used the protection of national security arguments in order to convince them do not publish the story about Merlin.

The victory of the official Washington did not last long. Three years later, in 2006, Risen published his book "State of War: The Secret History of the CIA and the Bush Administration". In the final Chapter 9 "A Rogue Operation", Risen not only disclosed Operation Merlin but also criticized it as mismanaged and actually aiding Iranian nuclear program instead of disrupting it. From his point of view, it was much easier finding flaws and fixing them in a completed design than starting from scratch. The government has continued arguing that Operation Merlin was an absolute success until its details were spread around by the journalist. Or maybe it was not such a success since shortly after Operation Merlin another secret operation against the Iranian nuclear program was launched. This time it was operation Olympic Games, which is better known by its main tool of sabotage- the computer worm Stuxnet.

By definition, sabotage was supposed to weaken the Iranian nuclear program, especially its uranium enrichment part. In reality, the pace of production of enriched uranium in Iran did not slow down. Maybe that was the reason for the computer worm Stuxnet to start as a cyber - weapon of sabotage, and later to be turned into the propaganda-weapon, but this is another story. The story of hunt and evolution...

Glossary

A

Air-gap. A protective measure that physically isolates computers from untrusted networks including the Internet.

Anti-virus software. A computer software designed to protect computers by detecting and removing malware.

Atomic Energy Commission. An agency of the United States government that was in charge of development of atomic science and technology during the period from 1946 to 1975.

B

BBC China. The German-owned cargo ship that was on its way from Dubai to Libya with a load of uranium enrichment equipment when it was diverted to the Italian port, Taranto. On October 4, 2003, CIA and MI6 seized the load (including the uranium enrichment centrifuges).

British Secret Intelligence Service (MI6). The foreign intelligence service of the United Kingdom.

C

Central Intelligence Agency (CIA). The foreign intelligence service of the United States of America.

Centrifugation. The uranium enrichment technology based on usage of centripetal force for separation of isotopes of Uranium-235 from isotopes of Uranium-238 based on their weight difference.

Computer worm. A malicious software with the ability to self-replicate in order to spread within a computer network from already infected computer to others.

D

Dichloroethane. A colorless liquid used as solvent and degreaser.

Dual-use equipment. Equipment that could be used for civilian and military purposes.

F

Firewall. A security system that protects computers by filtering communications between the computer and the Internet by allowing through only communication from trusted websites.

First Department. During the Soviet Union times, the First Department was responsible for secrecy of state organizations.

G

Gas centrifuge. A uranium enrichment centrifuge that works with uranium in its gaseous form – uranium hexafluoride.

Gaseous diffusion. The uranium enrichment technology. In the gaseous diffusion process, uranium hexafluoride is forced through a series of porous membranes. Under pressure, the light-weighted Uranium-235 moves through membranes faster than Uranium-238. As the result, it enriches uranium hexafluoride by increasing the concentration of Uranium-235.

Ghauri ballistic missile. A medium-range ballistic missile deployed by Pakistan Army.

H

Highly enriched uranium. Uranium with 20% or more concentration of Uranium-235.

I

India-Pakistan Winter war. The military conflict between India and Pakistan that took place from December 3 to December 16, 1971. The war ended with the defeat of Pakistani army and liberation of East Pakistan, which became a new independent state – Bangladesh.

Institute "A". A research institute that was located near Sukhumi in Georgia, the Soviet Union. After World War II, it was repurposed from a sanatorium to a research facility for German scientists from Soviet war-prisoners camps.

International Atomic Energy Agency (IAEA). An international organization in charge of nuclear non-proliferation with headquarters in Vienna, Austria.

K

Khan's network. An international nuclear proliferation network organized by Abdul Qadeer Khan.

L

Low-enriched uranium. Uranium with less than 20% concentration of Uranium-235.

M

Malware. A short form for malicious software that has been designed with an intent to operate against computer users and their needs.

N

Natanz uranium enrichment plant. A uranium enrichment facility in Iran that uses uranium enrichment centrifuges. Natanz was a main target of the Stuxnet computer worm.

Network. A cable or wireless connection between several computers in order to exchange data and share resources such as Internet access, printers, scanners, etc.

Nuclear non-proliferation. Attempts to stop the spread of nuclear weapons and related technology.

O

Operation Merlin. A covert operation by the CIA, which took place in 2000, aimed at supplying Iran with the altered blueprints of the Soviet-made nuclear weapon component - TBA-480 Fire Set. The alterations were supposed to make the component build according to the blueprints useless.

Operation Olympic Games. A covert operation aimed at sabotaging the Iranian nuclear program by using cyber-attacks. One of the cyber-weapons used during this operation was the computer worm Stuxnet.

P

P-1 and P-2 uranium centrifuges. Two first Pakistan-designed uranium centrifuges based on a Zippe-type centrifuge.

Pokhran Test Range. An Indian nuclear test site located in the Thar Desert (a.k.a. the Great Indian Desert) near the border with Pakistan.

R

Ras Koh Hills. The location of the mountain in the Chagai District of Balochistan. It was used by Pakistan for the detonation of five nuclear devices on May 28, 1998.

S

Sabotage. Deliberate efforts to interfere with production process by means of destroying or damaging equipment and/or raw materials and supplies.

Separative Work Unit (SWU). Unit of measure used to define the effort necessary for enrichment process of isotopes separation.

Smiling Buddha (a.k.a. Pokhran-I). The code name of the nuclear weapon test performed by India on May 18, 1974.

Stuxnet. A computer worm specifically created to attack and destroy uranium enrichment centrifuges at the Iranian enrichment plant.

T

TBA-480 Fire Set. The component of nuclear weapon designed in the Soviet Union, which is responsible for triggering an initial explosion necessary to start a nuclear chain reaction.

U

Uninterruptible power supply (UPS). An electrical device used as an emergency power source during unexpected interruption of electric power supply.

Uranium-234. An isotope of Uranium that is a decay product of Uranium-238. Its concentration in uranium ore is about 0.005%.

Uranium-235. An isotope of Uranium that is able to support a nuclear chain reaction. For this reason, it is suitable for use as fuel for nuclear power plants or as the main component of a nuclear warhead. In natural form, its concentration in uranium ore is less than 1%.

Uranium-238. An isotope of Uranium that represents more than 99% of uranium ore. When it comes to the production of nuclear fuel, Uranium-238 is useless, because it cannot support a nuclear chain reaction.

Uranium enrichment. A term used to describe a variety of processes aimed at increasing concentration of Uranium-235 isotope in processed uranium ore.

Uranium enrichment centrifuge. A gas centrifuge is an equipment that uses centripetal force for separation of isotopes of Uranium-235 from isotopes of Uranium-238 based on their weight difference.

Uranium hexafluoride. A chemical compound UF_6 that is used during the uranium enrichment process. Due to its physical properties, it could be converted to its gaseous form for processing, then turned to its liquid form for filling or emptying containers or equipment, and used in its solid form during storage. All that conversion could be done by changing temperature and pressure within a range commonly used in industrial processes.

Urenco Group. The international company that specializes in the production of nuclear fuel. It operates uranium enrichment facilities in Germany, the Netherlands, and the United Kingdom.

Z

Zippe-type uranium enrichment centrifuges. A gas centrifuge that uses centripetal force for separation of isotopes of Uranium-235 from isotopes of Uranium-238 based on their weight difference. Its design was patented by Gernot Zippe and Rudolf Scheffel in 1960.

Selected Bibliography

Chapter 1: The Backdoor to the Nuclear Powers Club

1. Bernstein, Jeremy. "A Nuclear Supermarket. (Physicists on Wall Street and Other Essays on Science and Society)" *SpringerLink*, Springer, New York, NY, 1 Jan. 1970. Web. 21 Nov. 2016. link.springer.com/chapter/10.1007/978-0-387-76506-8_7

2. Collins, Catherine, and Douglas Frantz. "Fallout: the True Story of the CIA's Secret War on Nuclear Trafficking". Free Press, 2014. https://www.amazon.com/Fallout-Story-Secret-Nuclear-Trafficking/dp/1439183066

3. Corera, Gordon. "Shopping for Bombs: Nuclear Proliferation, Global Insecurity,

and the Rise and Fall of the A.Q. Khan Network". Oxford University Press, 2009. https://www.amazon.com/Shopping-Bombs-Proliferation-Insecurity-Q/dp/0195375238/ref=mt_paperback?_encoding=UTF8&me=

4. Rehman, Shahidur. "Nuclear Proliferation: No AQ Khan Network, Only Western Suppliers." *The Express Tribune*, 6 Jan. 2011. Web. 20 Nov. 2016. tribune.com.pk/story/99782/nuclear-proliferation-no-aq-khan-network-only-western-suppliers/

5. Sanger, David E. "The Khan Network" Presented at the Conference on South Asia and the Nuclear Future held June 4-5, 2004 at Stanford University. Web. 21 Nov. 2016 fsi.stanford.edu/sites/default/files/evnts/media/Khan_network-paper.pdf

6. Implementation of the NPT Safeguards Agreement in the Islamic Republic of Iran (GOV /2006/27). *IAEA*. 28 April 2006. Web. 28 Sep. 2017. http://www.wsj.com/public/resources/documents/iranreport20060428.pdf

7. Implementation of the NPT Safeguards Agreement in the Islamic Republic of Iran (GOV /2003/75). *IAEA*. 10 Nov 2003. Web. 28 Sep. 2017.

https://www.iaea.org/sites/default/files/gov2003-75.pdf

Chapter 2: A. Q. Khan and the North Korean Nuclear Bomb

1. Corera, Gordon. "Shopping for Bombs: Nuclear Proliferation, Global Insecurity, and the Rise and Fall of the A.Q. Khan Network". Oxford University Press, 2009. https://www.amazon.com/Shopping-Bombs-proliferation-Insecurity-Q/dp/0195375238/ref=mt_paperback?_encoding=UTF8&me=

2. Henderson, Simon. "For the Love of Money". The Washington Institute for Near East Policy, 7 July 2011. Web. 2 Feb. 2017. www.washingtoninstitute.org/policy-analysis/pdf/for-the-love-of-money

3. Henderson, Simon. "Foreign Policy: Drunk with Power in North Korea." *NPR*. NPR, 08 July 2011. Web. 11 Feb. 2017. http://www.npr.org/2011/07/08/137696386/foreign-policy-drunk-with-power-in-north-korea

4. Filkins, Dexter. "N. Korea Aid to Pakistan Raises Nuclear Fears." *Los Angeles Times*, Los Angeles Times, 23 Aug. 1999. Web. 11 Feb. 2017. articles.latimes.com/1999/aug/23/news/mn-2949/2

5. Filkins, Dexter. "Who Killed Kim Sah Nae?" *The New Yorker*, The New Yorker, 20 June 2017. Web. Feb 11, 2017. http://www.newyorker.com/news/news-desk/who-killed-kim-sah-nae

6. Lintner, Bertil "Pyongyang's 60-year obsession." *Asia Times Online,* Asia Times Online, 10 Oct 2006. Web. Dec. 12, 2016. http://www.atimes.com/atimes/Korea/HJ10Dg02.html

7. "North Korea's Nuclear Tests in Pakistan." *Korea Web Weekly*, AFAR (Association for Asian Research), 30 Apr. 2004. Web. Dec. 12, 2016. www.asianresearch.org/articles/2019.html

8. Watson, Paul, and Mubashir Zaidi. "Death of N. Korean Woman Offers Clues to Pakistani Nuclear Deals." *Los Angeles Times,* Los Angeles Times, 1 Mar. 2004. Web. Feb. 12, 2017. articles.latimes.com/2004/mar/01/world/fg-murder1

Chapter 3: Gernot Zippe – the Creator of Uranium Enrichment Centrifuges

1. Broad, William J. "Slender and Elegant, It Fuels the Bomb." *The New York Times.* The New York Times, 22 Mar. 2004. Web. Nov. 12, 2016. http://www.nytimes.com/2004/03/23/science/slender-and-elegant-it-fuels-the-bomb.html?pagewanted=all&_r=0

2. Bukharin, Oleg. "Russia's Gaseous Centrifuge Technology And Uranium Enrichment Complex". *Program on Science and Global Security Woodrow Wilson School of Public and International Affairs.* Princeton University. January 2004. Web. 24 Oct. 2016. http://www.partnershipforglobalsecurity-archive.org/Documents/bukharinrussianenrichmentcomplexjan2004.pdf

3. "The History of the Gas Centrifuge and Its Role in Nuclear Proliferation." *Wilson Center,* 8 Nov. 2013. Web. 23 Oct. 2016. https://www.wilsoncenter.org/event/the-history-the-gas-centrifuge-and-its-role-nuclear-proliferation#sthash.e6BjJmQn.dpuf

4. Hoffmann, Dieter. "Fritz Lange, Klaus Fuchs, and the Remigration of Scientists to East Germany" *Springer.com*. Physics in Perspective, Dec. 2009. Web. 24 Oct. 2016.
http://download.springer.com/static/pdf/488/art%253A10.1007%252Fs00016-009-0427-5.pdf?originUrl=http%3A%2F%2Flink.springer.com%2Farticle%2F10.1007%2Fs00016-009-0427-5&token2=exp=1459175861~acl=%2Fstatic%2Fpdf%2F488%2Fart%25253A10.1007%25252Fs00016-009-0427-5.pdf%3ForiginUrl%3Dhttp%253A%252F%252Flink.springer.com%252Farticle%252F10.1007%252Fs00016-009-0427-5*~hmac=71d85a3ac20fc0fe2b3bd03ac891520d483a5ad608f25718c0b19e3e6a0b541d

5. Kestenbaum, David. "A History of the Centrifuge". *Morning Edition.* NPR. 21 Sep. 2005. Web. 25 Oct. 2016.
http://www.npr.org/templates/story/story.php?storyId=4857123

6. Zippe, Gernot. "Development and Status of Gas Centrifuge Technology", (accessed on 03/31/2014 at Orphaned (Re)Source: The Gernot Zippe Files) *Atomic Reporters*, Atomic Reporters. Web. 29 Oct. 2016. http://www.atomicreporters.com/wp-content/uploads/2015/10/zippe-story-1_upside.pdf

7. "The Zippe Type: Poor Man's Bomb." *BBC*. BBC Radio 4, 19 May 2004. Web. 02 Aug. 2017. http://www.bbc.co.uk/radio4/science/zippetype.shtml

8. Uranium Enrichment | Enrichment of Uranium. *World Nuclear Association.* World Nuclear Association. Web. 24 Oct. 2016. http://www.world-nuclear.org/information-library/nuclear-fuel-cycle/conversion-enrichment-and-fabrication/uranium-enrichment.aspx

9. Uranium Enrichment. *United States Nuclear Regulatory Commission.* United States Nuclear Regulatory Commission. Web. 24 Oct. 2016. http://www.nrc.gov/materials/fuel-cycle-fac/ur-enrichment.html

10. Wood, Houston G., Alexander Glaser, and R. Scott Kemp. "The Gas Centrifuge and Nuclear Weapons Proliferation." *Physics Today*, Physics Today, Sept. 2008. Web. 24 Oct. 2016.
http://ptonline.aip.org/journals/doc/PHTOAD-ft/vol_61/iss_9/40_1.shtml
Full text article in Physics Today:
http://scitation.aip.org/content/aip/magazine/physicstoday/article/61/9/10.1063/1.2982121

Chapter 4: Flashback: German Researchers in Soviet Captivity

1. Gorobetzh, Boris. "Three from the Atomic Project. The secret physicists Lejpunskie (Трое из Атомного проекта. Секретные физики Лейпунские)". Сетевой альманах ЕВРЕЙСКАЯ СТАРИНА 2007. Web. 26 Oct. 2016. http://berkovich-zametki.com/2007/Starina/Nomer4/Gorobec1.htm

2. Lowenhaupt, Henry S. "On the Soviet Nuclear Scent". *Center for the Study of Intelligence.* Central Intelligence Agency. 2 July 1996. Web, 6 March 2017.
https://www.cia.gov/library/center-for-the-study-of-intelligence/kent-csi/vol11no4/html/v11i4a02p_0001.htm

3. Oleynikov, Pavel. "German Scientists in the Soviet Atomic Project". *James Martin Center for Nonproliferation Studies.* Middlebury Institute of International Studies at Monterey. Summer 2000. Web. 28 Oct 2016. www.nonproliferation.org/wp-content/uploads/npr/72pavel.pdf

Chapter 5: Sabotage – the Weapon of Choice

1. Corera, Gordon. "Shopping for Bombs: Nuclear Proliferation, Global Insecurity, and the Rise and Fall of the A.Q. Khan Network". Oxford University Press, 2009. https://www.amazon.com/Shopping-Bombs-Proliferation-Insecurity-Q/dp/0195375238/ref=mt_paperback?_encoding=UTF8&me=

2. Collins, Catherine, and Douglas Frantz. "Fallout: the True Story of the CIA's Secret War on Nuclear Trafficking". Free Press, 2014. https://www.amazon.com/Fallout-Story-Secret-Nuclear-Trafficking/dp/1439183066

3. Maas, Peter. "CIA's Jeffrey Sterling Sentenced to 42 Months for Leaking to New York Times Journalist". The Intercept. May 11, 2015.Web, May 12, 2015. https://theintercept.com/2015/05/11/sterling-sentenced-for-cia-leak-to-nyt/

4. Risen, James. "State of War: The Secret History of the CIA and the Bush Administration". Free Press, 2007. https://www.amazon.com/State-War-Secret-History-Administration/dp/0743270673/ref=sr_1_1?ie=UTF8&qid=1509416100&sr=8-1&keywords=James+Risen.+State+of+War%3A+The+Secret+History+of+the+CIA+and+the+Bush+Administration.

5. Zetter, Kim. *Countdown to Zero Day: Stuxnet and the Launch of the World's First Digital Weapon.* Broadway Books, 2014. https://www.amazon.com/Countdown-Zero-Day-Stuxnet-Digital/dp/0770436196/ref=sr_1_4?s=books&ie=UTF8&qid=1510523619&sr=1-4&keywords=zero+day&dpID=61XdPxNxKdL&preST=_SY291_BO1,204,203,200_QL40_&dpSrc=srch

www.ingramcontent.com/pod-product-compliance
Lightning Source LLC
Chambersburg PA
CBHW030454220526
45464CB00006B/2527